Intratheater Airlift Functional Area Analysis (FAA)

T0195510

David T. Orletsky, Anthony D. Rosello, John Stillion

Prepared for the United States Air Force
Approved for public release; distribution unlimited

PROJECT AIR FORCE

The research described in this report was sponsored by the United States Air Force under Contract FA7014-06-C-0001. Further information may be obtained from the Strategic Planning Division, Directorate of Plans, Hq USAF.

Library of Congress Cataloging-in-Publication Data

Orletsky, David T., 1963-
    Intratheater airlift functional area analysis (FAA) / David T. Orletsky,
Anthony D. Rosello, John Stillion.
        p. cm.
    Includes bibliographical references.
    ISBN 978-0-8330-4417-4 (pbk. : alk. paper)
    1. Airlift, Military—United States—Planning. 2. Transportation, Military—
United States. 3. United States. Air Force—Transportation. I. Rosello, Anthony D.
II. Stillion, John. III. Title.

UC333.O75 2011
358.4'11—dc22

                                                                            2009026697

The RAND Corporation is a nonprofit institution that helps improve policy and decisionmaking through research and analysis. RAND's publications do not necessarily reflect the opinions of its research clients and sponsors.

RAND® is a registered trademark.

Published 2011 by the RAND Corporation
1776 Main Street, P.O. Box 2138, Santa Monica, CA 90407-2138
1200 South Hayes Street, Arlington, VA 22202-5050
4570 Fifth Avenue, Suite 600, Pittsburgh, PA 15213-2665
RAND URL: http://www.rand.org/
To order RAND documents or to obtain additional information, contact
Distribution Services: Telephone: (310) 451-7002;
Fax: (310) 451-6915; Email: order@rand.org

Preface

This functional area analysis (FAA) for U.S. Air Force intratheater airlift is the first in a series of documents RAND Project AIR FORCE is producing that together constitute a capabilities-based assessment (CBA) required as part of the Joint Capabilities Integration and Development System (JCIDS). According to Chairman of the Joint Chiefs of Staff Instruction (CJCSI) 3170.01E, an FAA identifies the operational tasks, conditions, and standards needed to achieve military objectives.[1] Two other elements of the JCIDS process will follow: the functional needs analysis (FNA) and the functional solution analysis (FSA).[2] The research described in this monograph was sponsored by Maj Gen Thomas P. Kane, Director, Plans and Programs, Headquarters, Air Mobility Command (AMC), Scott Air Force Base, Illinois (HQ AMC/A5). The work was conducted within the Force Modernization and Employment Program of RAND Project AIR FORCE as part of a fiscal year 2006 study, "Improving Air-Ground Integration, Interoperability, and Interdependence."

---

[1] CJCSI 3170.01E, *Joint Capabilities Integration and Development System*, May 11, 2005, p. A-4.

[2] John Stillion, David T. Orletsky, and Anthony D. Rosello, *Intratheater Airlift Functional Needs Analysis (FNA)*, Santa Monica, Calif.: RAND Corporation, MG-822-AF, 2011, and David T. Orletsky, Daniel M. Norton, Anthony D. Rosello, William Stanley, Michael Kennedy, Michael Boito, Brian G. Chow, and Yool Kim, *Intratheater Airlift Functional Solution Analysis (FSA)*, Santa Monica, Calif.: RAND Corporation, MG-818-AF, 2011.

RAND Project AIR FORCE

RAND Project AIR FORCE (PAF), a division of the RAND Corporation, is the U.S. Air Force's federally funded research and development center for studies and analyses. PAF provides the Air Force with independent analyses of policy alternatives affecting the development, employment, combat readiness, and support of current and future aerospace forces. Research is conducted in four programs: Force Modernization and Employment; Manpower, Personnel, and Training; Resource Management; and Strategy and Doctrine.

Additional information about PAF is available on our website:
http://www.rand.org/paf/

Contents

Figure and Tables

Figure

Tables

Summary

JCIDS implements a CBA and establishes a set of procedures for it to

identify, assess, and prioritize joint military capability needs.[1]

The system requires a series of analyses to identify capabilities gaps and to evaluate materiel and nonmateriel approaches to closing the gap. The FAA is the first in this series; it

identifies the operational tasks, conditions, and standards needed to achieve military objectives.[2]

The second in the series, the FNA,

assesses the ability of the current and programmed warfighting systems to deliver the capabilities the FAA identified under the full range of operating conditions and to the designated measures of effectiveness.[3]

The last item is the FSA. It

is an operationally based assessment of all potential [doctrine, organization, training, materiel, leadership and education, per-

---

[1] Department of Defense Instruction 5000.2, *Operation of the Defense Acquisition System*, May 12, 2003; CJCSI 3170.01E, 2005, p. 1.

[2] CJCSI 3170.01E, 2005, p. A-4.

[3] CJCSI 3170.01E, 2005, p. A-4.

sonnel, and facilities] and policy approaches to solving (or mitigating) one or more of the capability gaps identified in the FNA.[4]

The broad objective of these three documents is to determine whether specific shortfalls in military capabilities require materiel solutions or whether modifying other aspects of the system could resolve the shortfall.

This CBA was initiated to analyze a potential deficiency in intratheater airlift capability. The Air Force identified three broad operational mission areas relating to the intratheater airlift system for this evaluation,[5] centering it on the system's ability to provide

1. *routine sustainment,* defined as the steady-state delivery of required supplies and personnel to units
2. *time-sensitive, mission-critical resupply,* defined as the delivery of supplies and personnel on short notice, outside the steady-state demands
3. *maneuver* to U.S. and allied forces across all operating environments, defined as the transport of combat teams around the battlefield using the intratheater airlift system.[6]

The JCIDS process requires the CBA to start with high-level guidance from the National Security Strategy and the National Defense Strategy. Individual service concepts of operation and the Family of Joint Future Concepts, both developed from the national strategies, also inform the process. We used these documents for input and guidance so that we could "identify tasks, conditions, and standards" required for the intratheater airlift fleet.[7] We also considered recent experience in Afghanistan and Iraq for insight into the current operational environment. (See pp. 32–33.)

---

[4]   CJCSI 3170.01E, 2005, p. A-4.

[5]   Meeting at Air Mobility Command, December 8, 2005, and subsequent discussions with Air Force personnel.

[6]   Meeting at Air Mobility Command, December 8, 2005, and subsequent discussions with Air Force personnel.

[7]   CJCSI 3170.01E, 2005, p. A-4.

We developed sets of tasks, conditions, and standards considered important for this CBA. Table S.1 presents the tasks derived during the FAA and identifies their applicability to each of the three mission areas discussed.

Although the guidance documents do not specify a set of conditions under which these tasks *must* be accomplished, attributes and conditions are discussed throughout the guidance documents. Some of these attributes and/or conditions occur in multiple guidance documents. The following conditions were deemed important (see p. 17):[8]

- adverse weather
- multiple, simultaneous, distributed decentralized battles and campaigns
- antiaccess environment

Table S.1
Tasks and Mission Areas Applicable to This CBA

Task	Routine Sustainment	Time-Sensitive, Mission-Critical Resupply	Small-Unit Maneuver
Transport supplies and equipment to points of need	X	X	X
Conduct retrograde of supplies and equipment	X	X	X
Transport forces and accompanying supplies to points of need[a]			X
Conduct recovery of personnel and supplies[b]			X
Transport replacement and augmentation personnel	X	X	X
Evacuate casualties	X	X	X

[a] Deployment, redeployment, and retrograde.

[b] Including evacuation of hostages, evacuees, enemy personnel, and high-value items.

---

[8] This and the following lists were compiled from multiple guidance documents. Much of the language derives directly from them.

- in support of forces operating in and from austere or unimproved locations
- degraded environment (weapons of mass destruction or effect; chemical, biological, radiological, nuclear, and explosive weapons; natural disasters)
- multinational environment
- absence of preexisting arrangement
- consistent with sea basing.

The guidance documents describe the following attributes and conditions as positive (see pp. 18–20):

- smallest logistical footprint
- speed, accuracy, and efficiency
- distribution to the point of requirement
- basing flexibility—the ability to operate across strategic and operational distances.

The documents also specify standards for evaluating potential gaps in capabilities. The tasks identified above should be accomplished with the following standards in mind (see pp. 14–15):

- ability to meet force and materiel movement demand
- ability to deliver optimized movement of forces and materiel throughout theater from a cycle time perspective
- capability to provide materiel support for the current and planned operations.

The results of the analysis will be sensitive to variables describing the operational environment and operational tasks. Some of the more-important variables are the number of delivery points; the terrain; air base accessibility; the total amount, size, and weight of each supply class to be delivered by air; the number of personnel to be delivered by air; the required response time; and the threat level.

Acknowledgments

The authors are grateful to many individuals in the U.S. Air Force. Maj Gen Thomas Kane, Director of Strategic Plans and Programs, Headquarters AMC, was the sponsor of this work. At AMC, Maj Paul Jones, Maj Vernon Lucas, Maj Mark Hering, Maj Ed Koharik, Maj Bill Spangenthal, Maj Dave Herbison, and Mike Houston provided a great deal of insight and assistance throughout the project. Air Force officers on the Air Staff also provided significant assistance. Maj Frank Altieri and Maj Jefe Brown were instrumental.

At RAND, we wish to thank David Shlapak and John Halliday for their helpful reviews of an earlier draft.

Abbreviations

AMC	Air Mobility Command
AMMP	Air Mobility Master Plan
CBA	capabilities-based assessment
CBRNE	chemical, biological, radiological, nuclear, and explosive
CCJO	Capstone Concept for Joint Operations
CJCSI	Chairman of the Joint Chiefs of Staff Instruction
CJCSM	Chairman of the Joint Chiefs of Staff Manual
CONOPS	concepts of operation
CS	civil support
DoD	Department of Defense
EP	emergency preparedness
FAA	functional area analysis
FNA	functional needs analysis
FSA	functional solution analysis
HLD	Homeland Defense
HLS	Homeland Security

JCIDS	Joint Capabilities Integration and Development System
JDDE	Joint Deployment and Distribution Enterprise
JFC	Joint Functional Concept
JIC	Joint Integrating Concept
JOC	Joint Operating Concept
JOpsC	Joint Operations Concept
JROC	Joint Requirements Oversight Council
MAF	Mobility Air Forces
MANPADS	man-portable air defense system
MCL	Master Capabilities Library
NDS	National Defense Strategy
NSS	National Security Strategy
OF	objective force
OP	operational-level (UJTL)
PAF	RAND Project AIR FORCE
SOF	special operations forces
TA	tactical-level (UJTL)
TS/MC	time-sensitive, mission-critical
UJTL	Universal Joint Task List
USAF	U.S. Air Force
WMD	weapons of mass destruction
WME	weapons of mass effect

CHAPTER ONE

Introduction, Purpose, and Scope

This functional area analysis (FAA) is the first in a series of documents that together constitute a capabilities-based assessment (CBA) as required by the Joint Capabilities Integration and Development System (JCIDS).[1] According to Chairman of the Joint Chiefs of Staff Instruction (CJCSI) 3170.01E, the FAA

> identifies the operational tasks, conditions, and standards needed to achieve military objectives.[2]

The instruction further states that the functional needs analysis (FNA), which follows the FAA,

> assesses the ability of the current and programmed warfighting systems to deliver the capabilities the FAA identified under the full range of operating conditions and to the designated measures of effectiveness.[3]

Finally, it describes the functional solution analysis (FSA) as

---

[1] Department of Defense Instruction 5000.2, *Operation of the Defense Acquisition System*, May 12, 2003. Note that the Department of Defense (DoD) updated this instruction in 2008, well after we completed the groundwork for our analysis.

[2] CJCSI 3170.01E, *Joint Capabilities Integration and Development System*, May 11, 2005, p. A-4. Since we began our work, this instruction has been revised twice. Much of the material describing the CBA process and the F-series documents has been split off into a second document: Chairman of the Joint Chiefs of Staff Manual (CJCSM) 3170.01C, *Operation of the Joint Capabilities Integration and Development System*, May 1, 2007.

[3] CJCSI 3170.01E, 2005, p. A-4.

an operationally based assessment of all potential [doctrine, organization, training, materiel, leadership and education, personnel, and facilities] and policy approaches to solving (or mitigating) one or more of the capability gaps identified in the FNA.[4]

The broad objective of this "F-series" of documents is to determine whether a materiel solution is required to address specific shortfalls in military capabilities or whether modifying other aspects of the system could resolve the shortfall.

The purpose of this specific CBA is to analyze a potential deficiency of intratheater airlift capability within the Mobility Air Forces (MAF).[5] The CBA was prompted by a concern that demands from the ongoing global war on terrorism and new operational concepts might lead to a shortfall in the ability of the U.S. Air Force (USAF) to deliver personnel and equipment.

Traditional airlift of personnel and materiel, via airdrop or airland methods, has been employed on numerous occasions during recent operations when surface lines of communication were not accessible or when cargo needed to arrive quickly. Effective support of current and future ground combat operations may require capabilities that do not exist in the current programmed USAF airlift fleet. For example, capabilities for operation from short and rough fields or for aircraft survivability required to support future Army CONOPS could be well beyond the capabilities of the current USAF intratheater airlift fleet.

This analysis, which will result in the F-series of documents, was initiated primarily to investigate the possibility that a light cargo airlift—that is, an aircraft with a smaller payload capacity and the ability to use shorter fields than a C-130—could be a cost-effective means of bridging potential intratheater airlift shortfalls. This class of aircraft, proponents suggest, would enable a more-effective and more-efficient

---

[4] CJCSI 3170.01E, 2005, p. A-4.

[5] Although we consider some attributes of airlift that have been traditionally thought of as belonging to special operations forces (SOF), our analysis focuses exclusively on Air Mobility Command's (AMC's) conventional airlift aircraft. This FAA discusses SOF missions because future Army concepts of operation (CONOPS) may become more like special operations. AMC should be aware of the resulting potential change in intratheater airlift operations.

intratheater airlift system than currently exists and improve support of ground combat forces. However, to avoid presupposing an airlift solution, delivery method, or even a materiel solution, the FAA will describe the desired capabilities only in general terms. This generality will allow for proper exploration and analysis to define the needed capability precisely and is in the spirit of the JCIDS process, allowing the exploration of both materiel and nonmateriel solutions.

This assessment focuses on the intratheater cargo and personnel delivery mission. This mission is primarily driven by the joint land force requirement to move personnel, equipment, and supplies throughout the battlespace. The Army has already completed its own FAA, FNA, and FSA on fixed-wing aviation, which identified several shortfalls requiring a materiel solution.[6] The 2005 Quadrennial Defense Review required a joint program office to implement the acquisition of any aircraft procured as a result of the Army studies.[7] In February 2006, the chiefs of staff of the USAF and Army signed a memorandum of understanding that directed the services to develop a joint memorandum of agreement within 90 days to articulate the path forward for each of the services toward developing complementary capabilities with respect to light cargo aircraft.[8] As a result, this FAA was undertaken as the jumping-off point for USAF involvement in this joint program.

---

[6] U.S. Army Aviation Center, Futures Development Division, Directorate of Combat Developments, *Army Fixed Wing Aviation Functional Area Analysis Report*, Fort Rucker, Ala., June 3, 2003a; U.S. Army Aviation Center, Futures Development Division, Directorate of Combat Developments, *Army Fixed Wing Aviation Functional Needs Analysis Report*, Fort Rucker, Ala., June 23, 2003b; and U.S. Army Aviation Center, Futures Development Division, Directorate of Combat Developments, *Army Fixed Wing Aviation Functional Solution Analysis Report*, Fort Rucker, Ala., June 8, 2004.

[7] DoD, *Quadrennial Defense Review Report*, February 6, 2006. See also U.S. Army, Training and Doctrine Command Analysis Center, *Future Cargo Aircraft (FCA) Analysis of Alternatives (AoA)*, July 18, 2005, not releasable to the general public.

[8] U.S. Army and USAF, "Way Ahead for Convergence of Complementary Capabilities," memorandum of understanding, February 2006.

The USAF identified three broad operational mission areas relating to the intratheater airlift system that should be evaluated in this analysis.[9] These areas are the system's ability to provide

1. routine sustainment
2. time-sensitive, mission-critical (TS/MC) resupply
3. maneuver

to U.S. and allied forces across all operating environments.

Routine sustainment is defined as the steady-state delivery of required supplies and personnel to units. *TS/MC resupply* is defined as the delivery of supplies and personnel on short notice, in addition to regular steady-state demands. *Maneuver* is defined as the transport of combat teams around the battlefield using the intratheater airlift system. These three operational mission areas have different characteristics and impose different requirements on the system. Each will be analyzed using a different but related construct. These operational mission areas are further defined later in this document. The capabilities and tasks listed in this FAA are required of the global mobility system within the framework of national, joint, USAF, and Army operational concepts.

Chapter Two provides a description of the JCIDS process and relevant national security concept and strategy documents. Chapter Three describes the operational mission areas. Chapter Four discusses the scenarios and operational environment, and Chapter Five describes the operational tasks derived from the Universal Joint Task List (UJTL) and the Air Force Master Capabilities Library (MCL). The final chapter presents our concluding remarks on identifying the tasks, conditions, and standards for intratheater airlift.

---

[9] Meeting at AMC, December 8, 2005, and subsequent discussions with USAF personnel.

CHAPTER TWO

Guidance Documents

JCIDS Process Guidance

CJCSI 3170.01E establishes the "policies and procedures of JCIDS"[1] as specified in the U.S. Code. JCIDS and its validated and approved documentation provide the Chairman of the Joint Chiefs of Staff "advice and assessment in support"[2] of the laws governing military acquisition. CJCSI 3170.01E

> also provides joint policy, guidance and procedures for recommending changes to existing joint resources when these changes are not associated with a new defense acquisition program.[3]

CJCSI 3170.01E provides a top-down process to identify needed capabilities. The process begins with high-level guidance from the National Security Strategy (NSS) and the National Defense Strategy (NDS). Individual service CONOPS and the Family of Joint Future Concepts, both developed from the national strategies, also inform and may initiate a need for new capabilities.[4]

Once a potential new capability is identified, JCIDS vets it through a standardized analysis process. The results of the analysis process are then used to make recommendations on how best to

---

[1] CJCSI 3170.01E, 2005, p. 1.

[2] CJCSI 3170.01E, 2005, p. 1.

[3] CJCSI 3170.01E, 2005, p. 1.

[4] This text was drawn from CJCSI 3170.01E, 2005, p. A-4, Figure A-1.

acquire the needed capability through possible changes in doctrine, organization, training, materiel, leadership and education, personnel, and facilities (DOTMLPF) and policy.

Sources and Guidance

The FAA uses multiple sources for input and guidance. The following paragraphs briefly describe the documents that informed this FAA.

National Security Strategy

The 2002 NSS articulates, at the highest level, our national goals for security and specific avenues for achieving them:[5]

> political and economic freedom, peaceful relations with other states, and respect for human dignity. . . .

> To achieve these goals, the United States will:

- champion aspirations for human dignity;
- strengthen alliances to defeat global terrorism and work to prevent attacks against us and our friends;
- work with others to defuse regional conflicts;
- prevent our enemies from threatening us, our allies, and our friends, with weapons of mass destruction;
- ignite a new era of global economic growth through free markets and free trade;
- expand the circle of development by opening societies and building the infrastructure of democracy;
- develop agendas for cooperative action with other main centers of global power; and
- transform America's national security institutions to meet the challenges and opportunities of the twenty-first century.[6]

---

[5]   George W. Bush, *The National Security Strategy of the United States of America*, Washington, D.C.: The White House, September 17, 2002.

[6]   Bush, 2002, pp. 1–2.

National Defense Strategy

The NDS provides DoD's approach to supporting the NSS and to meeting future defense challenges.[7] The strategy outlines "an active, layered approach to the defense of the nation and its interests."[8] The NDS lists a series of strategic objectives:

- Secure the United States from direct attack. . . .
- Secure strategic access and retain global freedom of action. . . .
- Strengthen alliances and partnerships. . . .
- Establish favorable security conditions. . . .[9]

These strategic objectives are accomplished by assuring allies and friends, dissuading potential adversaries, deterring aggression, countering coercion, and defeating adversaries.[10] Four guidelines structure the strategic planning and decisionmaking:

- Active, layered defense. . . .
- Continuous transformation. . . .
- Capabilities-based approach. . . .
- Managing risk. . . . [11]

Joint Operations Concepts Family of Documents

In April 2003,

> the Secretary of Defense directed the development of the Joint Operations Concepts (JOpsC) family. This family consists of a Capstone Concept for Joint Operations (CCJO), Joint Operating Concepts (JOCs), Joint Functional Concepts (JFCs), and Joint

---

[7]   DoD, *The National Defense Strategy of the United States of America*, March 2005a. DoD produced a new version of this document in mid-2008, well after we completed groundwork for this volume.

[8]   DoD, 2005a, p. iv.

[9]   DoD, 2005a, p. iv.

[10]   DoD, 2005a, p. iv.

[11]   DoD, 2005a, p. iv.

Integrating Concepts (JICs). These concepts look beyond the Future Years Defense Plan out to 20 years.[12]

This family of documents continues to evolve. Some documents are in draft form, others in final form, and some forthcoming. This discussion uses the documents available at the time of writing.[13]

The CCJO is the "overarching concept of the family of joint concepts." It "broadly describes how joint forces will operate" in the future environment, which it specifies as being 2012 to 2025.[14] JOCs describe, at the operational level, how a joint force commander will "accomplish a strategic mission through the conduct of operational-level military operations within a campaign."[15] JOCs apply

> the CCJO solution and joint force characteristics to a more specific military problem. JOCs also identify challenges, key ideas for solving those challenges, effects to be generated to achieve objectives, essential capabilities likely needed to achieve objectives and the relevant conditions in which the capabilities must be applied.[16]

Additionally, JOCs affect development of Defense Planning Scenarios and drive development of service and joint transformation road maps.

JFCs describe how the future joint forces will perform a particular military function across the full range of military operations. JFCs apply the CCJO solution and joint force characteristics to the specific military problem and identify the required functional capabilities needed to generate the effects described in JOCs and identify attributes needed to functionally support the future joint force.

---

[12] Joint Staff, Joint Experimentation, Transformation, and Concepts Division (J7), "JOpsC Family of Joint Concepts—Executive Summaries," briefing, August 23, 2005b, slide 3.

[13] Joint Staff (J7), "Joint Integrating Concepts," *Future Joint Warfare* website, last update January 2, 2008.

[14] Joint Staff (J7), 2005b, slide 5.

[15] Joint Staff (J7), 2005b, slide 6.

[16] Joint Staff (J7), 2005b, slide 6.

JICs describe how a joint force commander will perform his operations or functions that are a subset of JOC and JFC capabilities. JICs have the narrowest focus of the JOpsC family. The JICs describe capabilities, decompose them into task-level detail, and include an illustrative vignette of an operating environment as a backdrop for the required tasks.[17]

In the case of a JROC-directed CBA, a CBA-specific JIC is typically produced: "To date, all Joint Requirements Oversight Council (JROC)–directed CBAs have been accompanied by a JIC."[18] Since this CBA is not JROC-directed and has a fairly narrow focus, a CBA-specific JIC does not exist. We therefore looked to existing JICs for guidance.

Figure 2.1 presents the JOpsC family of documents graphically. Under each heading is a list of the documents currently available. The documents that are listed in bold are the ones most relevant to this CBA. Each of these is discussed in more detail below.

The JOCs currently in draft or final form include Homeland Defense (HLD) and Civil Support,[19] Strategic Deterrence, Major Combat Operations, and Stability Operations. JFCs include Battlespace Awareness, Command and Control, Force Application, Focused Logistics, Force Protection, Net-Centric Operations, Force Management, and Training. JICs include Global Strike, Joint Forcible Entry Operations, Joint Undersea Superiority, Integrated Air and Missile Defense, Seabasing, Joint Logistics, and the Joint Command and Control.[20] The CCJO and all JOCs impact this CBA. Further, all JFCs and JICs could

---

[17] Edward Yarnell, "Joint Transformation Concepts," briefing, Joint Experimentation, Transformation and Concepts Division (J7), January 6, 2006.

[18] Joint Staff, Force Application Assessment Division (J8), "Conducting a Capabilities-Based Assessment (CBA) Under the Joint Capabilities Integration and Development System (JCIDS)," white paper, January 2006, p. 11.

[19] DoD, *Homeland Defense and Civil Support Joint Operating Concept*, draft, Vers. 1.5, November 2005c.

[20] This discussion of currently available JOCs, JFCs, and JICs was primarily derived from a briefing (Joint Staff [J7], 2005b). In our discussion, some of the titles are slightly different from those in the briefing because the authors were working with a subsequent version of the particular document with a different name.

Figure 2.1
JOpsC Family of Documents

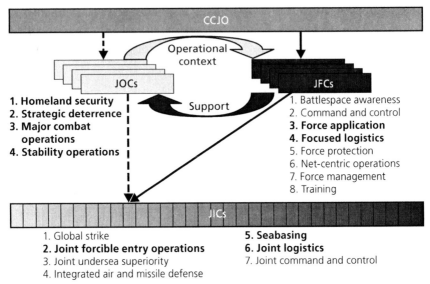

SOURCE: Adapted from Joint Staff (J7), 2005b.
RAND *MG685-2.1*

be of importance—especially as these documents are updated. Specific concepts that are especially relevant to this FAA include the Focused Logistics JFC and the Joint Logistics JIC.

The CCJO's solution section is particularly relevant to this FAA. Specifically, the fundamental action to "establish, expand and secure reach"[21] relates directly to the routine sustainment, TS/MC resupply, and maneuver of this FAA.

The Major Combat Operations JOC describes the foundations for major combat operations and how the joint force fights.[22] Capabilities in this JOC that are applicable to our FAA include Force Application and Focused Logistics.[23] Relevant Force Application capabilities involve the ability to "rapidly project" and "sustain" forces "through-

---

[21] Joint Staff (J7), 2005b, p. 12.

[22] DoD, *Major Combat Operations Joint Operating Concept*, September 2004e.

[23] DoD, 2004e, pp. 56–58.

out the battlespace" while "eliminating redundancies" and "enhancing the effectiveness."[24] The essence of Focused Logistic capabilities is best captured in the following capability from the Major Combat Operations JOC:

> Project and sustain forces when the adversary is competent and determined, strategic and theater lines of communication are not secure, access through fixed seaports and airfields is denied and supported forces are widely dispersed in the battlespace.[25]

The Homeland Security JOC describes how "DoD intends to fulfill its responsibilities associated with securing the Homeland, to include HLD, CS [Civil Support] and Emergency Preparedness (EP)."[26] It presents

> broad operational-level objectives, scopes the depth and breadth of HLD and CS operations and EP responsibilities, and outlines how DoD will accomplish them.[27]

---

[24] DoD, 2004e, p. 57.

[25] DoD, 2004e, p. 58, 4D.4.

[26] DoD, 2005c, p. ES-1. Note that this document is Version 1.5, a draft of the second iteration of the Homeland Security (HLS) JOC, Vers. 1.0, February 2004. The draft expands the scope slightly, but the characteristics required to accomplish the mission remain unchanged from the HLS JOC. Although we discuss the HLD and CS JOC, we believe that these concepts are less relevant to this CBA than the others are. First, as discussed earlier, this CBA was undertaken to explore potential shortfalls experienced during the current operations outside the United States—operations in Iraq and Afghanistan—not homeland security. Second, the potential shortfalls for intratheater airlift assets often relate to the capability to use short, austere airfields, which is less of an issue in the United States. Most needs in the United States proper will likely arise in areas with large population densities, which almost invariably have advanced and robust aviation infrastructures. Third, the entire U.S. transportation system is enormously advanced relative to other regions of the world. Many HLD and CS situations that one can envision would best be handled using surface transportation.

[27] DoD, 2005c, p. ES-4.

Essential characteristics to accomplish the missions and objectives include forces that are "Fully Integrated, Expeditionary, Networked, Decentralized, Adaptable, Decision Superior, and Effective."[28]

The Stability Operations JOC

> articulates how a future joint force commander plans, prepares, deploys, employs, and sustains a joint force conducting stability operations that precede, occur during and follow conventional combat operations.[29]

Stability Operations JOC capabilities that may be relevant to this CBA fall under Focused Logistics:

- Support restoration of basic services by identifying those needs during planning and throughout the execution phase.[30]
- The ability to account for, contain, distribute, or destroy military spoils: weapons, ammunition, and equipment and to conduct sensitive weapon site preservation.[31]
- The ability to rapidly provide essential civil assistance, humanitarian, and reconstruction materiel in a combat or other hostile environment.[32]

In addition, two capabilities can be drawn from the Force Application section:

- The ability to integrate deployment, employment and sustainment of the force, thus eliminating redundancies, stimulating synergy, and coordinating the movement and sustainment of forces conducting stability operations, and reducing in-country and regional footprint.[33]

---

[28] DoD, *Force Application Functional Concept*, March 5, 2004c, pp. 43–45.

[29] DoD, *Stability Operations Joint Operating Concept*, September 2004d, p. 1.

[30] DoD, 2004d, p. 51.

[31] DoD, 2004d, p. 51.

[32] DoD, 2004d, p. 51.

[33] DoD, 2004d, p. 50.

- The ability to seamlessly transition joint deployment, employment and sustainment from supporting preventive transition actions, to being supported during major conventional combat operations, and then supporting post-combat operations.[34]

The Strategic Deterrence JOC

describes how the Joint Force Commanders . . . will plan, prepare, deploy, employ, and sustain a joint force to contribute to a strategic deterrence strategy set forth by national leadership through 2015.[35]

This JOC is the least important to this CBA but is described here for completeness.

Relevant JFCs and JICs

JFCs

The Focused Logistics and Force Application JFCs are most applicable to this CBA.

The following challenges are from Focused Logistics:

- Joint Deployment/Rapid Distribution
- Agile Sustainment
- Operational Engineering
- Multinational Logistics
- Force Health Protection
- Joint Theater Logistics Management.[36]

---

[34] DoD, 2004d, p. 51.

[35] DoD, *Strategic Deterrence Joint Operating Concept*, February 2004b, p. 3.

[36] DoD, *Focused Logistics Joint Functional Concept*, Vers. 1.0, December 2003b, p. 10-13.

Capabilities essential for meeting the challenges that are most relevant to this CBA are discussed below.[37]

The capabilities essential for meeting the joint deployment/rapid distribution challenge include the following:

- A fully enabled mobility system, with full-spectrum-capable mobility forces in the right numbers and types, supported by a robust infrastructure, and further characterized by capabilities.
  - optimizing rapid projection, delivery, and handoff of joint forces and sustainment assets worldwide
  - distributing required forces and sustainment at the place and time required
  - supporting rapid force maneuver within the joint or combined operations area
  - returning forces to the sea base, home station, or other location for regeneration and reconstitution.
- Effective and efficient deployment and distribution process . . .
  - Integrated vertically and horizontally from the strategic to the tactical level . . . .[38]

One capability "essential for meeting the agile sustainment challenge" is "[p]recision tactical resupply, including—but not limited to—delivery by airdrop, precision aerial delivery, or airland."[39]

The JROC-approved attributes for Focused Logistics are[40]

- Fully Integrated
- Expeditionary

---

[37] DoD, 2003b, pp. 22–26, lists other requirements. We have taken note of the other capabilities and will keep them in mind throughout the analysis. We present here only a subset of the identified capabilities that are directly related to this CBA.

[38] DoD, 2003b, pp. 22–23.

[39] DoD, 2003b, p. 23.

[40] The phrase "JROC-approved attributes" was taken from Joint Chiefs of Staff, *Focused Logistics Campaign Plan*, 2004b, p. 20. The JFC lists the same attributes on p. 26 but does not identify them as JROC-approved.

- Networked
- Decentralized
- Adaptable
- Decision Superiority.[41]

In addition, the JFC identifies several attributes for the "logistics capabilities . . . support function":

- Effective
- Reliable
- Affordable.[42]

It also identifies two metrics that are important for this CBA, which it describes as falling primarily under the "Warfighter Perspective":

- Force Movement: Capability to move forces and equipment to final destination in accordance with warfighter requirements.
  - Force Movement Capacity: Ability to meet force and materiel movement demand.
  - Force Movement Visibility & Control: Level of visibility of personnel, equipment, and supplies in the distribution network to support the warfighter demand.
  - Force Movement Effectiveness: Ability to deliver optimized movement of forces and materiel into theater from a cycle time perspective.[43]
- Force Sustainment: Capability to provide ongoing support for current and planned operations.
  - Materiel Support: Capability to provide materiel support for the current and planned operations.

---

[41] DoD, 2003b, pp. 26–27.

[42] DoD, 2003b, p. 27.

[43] In this context, we interpret *cycle time* as referring to the ability of an airlifter to turn sorties. That is, since we are assuming a continuous flow of airlifted materiel, the ability of an aircraft to make multiple deliveries during a day must be considered.

    — Services Support: Capability to provide services support for current and planned operations.[44]

The Force Application JFC "concentrates on those capabilities required to effectively apply force against large-scale enemy forces in the 2015 timeframe" and defines *force application* as the "integrated use of maneuver and engagement."[45] It further defines maneuver as the

> movement of forces into and through the battlespace to a position of advantage in order to generate or enable the generation of effects on the enemy.[46]

The Force Application JFC discusses both strategic and tactical agility. Agility is one of the 12 attributes this JFC identifies. It defines *strategic agility* as

> the ability to quickly move strategic distances and successfully conduct joint forcible entry into the theater of operations
>
> - with very short lead times
> - from long travel distances
> - with minimal infrastructure at intermediate staging locations
> - may need to operate from a joint sea base.[47]

The JFC defines *operational and tactical agility* as the ability to "operate at will within all domains in order to enable engagements across the depth and breadth of the battlespace." In some cases, clandestine maneuver is required. In all cases, the forces must be capable of moving "quickly in order to capitalize on fleeting tactical and operational opportunities."[48]

---

[44] DoD, 2003b, p. 29.

[45] DoD, 2004c, p. 4.

[46] DoD, 2004c, p. 10.

[47] DoD, 2004c, p. 10.

[48] DoD, 2004c, p. 11.

JICs

The JIC most applicable to this CBA is that for Joint Logistics (Distribution).[49] We drew on this one as a basis for required tasks; on the Joint Forcible Entry Operations JIC for some insight into the maneuver mission area; and, to some extent, on the Seabasing JIC. All these JICs are discussed in more detail below.

The Joint Logistics (Distribution) JIC describes tasks required to conduct the Joint Deployment and Distribution Enterprise (JDDE). This JIC presents 13 conditions that it describes as

directly affecting task performance:

- Adverse Weather (high sea state, low visibility, temperature extremes, etc.).
- Required Joint Reception, Staging, Onward Movement and Integration (JRSOI).
- Multiple, simultaneous, distributed decentralized battles and campaigns.
- Anti-access environment.
- Support forces operating in and from austere or unimproved locations.
- Degraded environments (WMD/WME [weapons of mass destruction or effect], CBRNE [chemical, biological, radiological, nuclear, and explosive], Natural disasters).
- Increased homeland security threat.
- Multi-national environment.
- Military culture supportive of JDDE.
- Authority and resources availability to enable JDDE initiatives.
- Completed time-phased force deployment data (TPFDD) available.
- Absence of pre-existing arrangement.
- Subject to JFC request for JDDE Support.[50]

The JIC goes on to state that the

---

[49] JCS, *Joint Logistics (Distribution) Joint Integrating Concept*, Vers. 1.0, February 7, 2006.

[50] JCS, 2006, pp. C-3 to C-4.

critical characteristics required of an effective and efficient JDDE are: Capacity, Visibility, Reliability, Velocity, and Precision.[51]

According to the Joint Logistics (Distribution) JIC, the JDDE has three major functions:

- Move the Joint Force
- Sustain the Joint Force
- Operate the JDDE.

"Move the Joint Force" is divided into *strategic* and *operational* movement. Strategic movement is beyond the scope of this CBA. Table 2.1 presents the tasks associated with operational movement. Table 2.2 presents the tasks associated with "Sustain the Joint Force."

The tasks associated with "Operate the JDDE" cover operating the logistical system as a whole and are not directly related to intratheater airlift. This CBA will address these tasks only when appropriate.

Although the Joint Forcible Entry Operations JIC that was available to us as we began our work was marked as a "draft for collaboration purposes only," it still offered some statements that may be applicable to this CBA:

- Establish the smallest logistical footprint but deliver with speed, accuracy, and efficiency.
- Eliminate strategic, operational, and tactical boundaries.
- Distribute to the point of requirement.
- Integrate a joint deployment, employment, and sustainment process which dynamically "senses" requirements and is adaptive and responsive.[52]
- Ensure freedom of movement for sustainment platforms.[53]

This document also discusses

---

[51] JCS, 2006, pp. C-4 to C-5.

[52] DoD, *Joint Forcible Entry Operations Joint Integrating Concept*, Vers. 92A3, limited distribution draft, September 15, 2004f, p. 37.

[53] DoD, 2004f, p. 37.

Table 2.1
Move the Joint Force Operational Tasks

Task Number	Joint Logistics (Distribution) JIC Task
1.2.1	Transport forces and accompanying supplies to points of need
1.2.1.1	Conduct onward movement operations
1.2.1.2	Conduct en route replenishment operations
1.2.2	Support reception and staging incident to intra-theater movement
1.2.2.1	Conduct reception operations
1.2.2.2	Conduct staging operations
1.2.3	Conduct intra-theater casualty movement
1.2.3.1	Conduct intra-theater patient movement
1.2.3.2	Conduct intra-theater movement of remains

SOURCE: Quoted from JCS, 2006, pp. C-11 to C-21.

Table 2.2
Sustain the Joint Force

Task Number	Joint Logistics (Distribution) JIC Task
2.1	Deliver supplies to the point of need
2.1.1	Position sustainment stocks
2.1.2	Cross-level sustainment
2.1.2.1	Deliver cross-leveled materiel to end user
2.1.2.2	Coordinate replenishment of cross-leveled materiel
2.1.3	Build tailored sustainment packages
2.2	Expand distribution capability to support global sustainment surge requirements
2.3	Conduct retrograde operations
2.3.1	Conduct retrograde of supplies
2.3.2	Conduct retrograde of equipment
2.4	Coordinate HNS [host-nation support], IA [interagency], MN [multinational], contractor, and nongovernmental organization distribution services; involves
2.4.1	Integrating performance-based logistic support activities
2.4.2	Coordinate direct vendor delivery
2.5	Deliver replacement/augmentation personnel

SOURCE: Quoted from JCS, 2006, pp. C-11 to C-21.

focused logistics capabilities:

- Deliver and sustain the joint forcible entry operations force, in all weather conditions, to objectives independent of existing infrastructure, from remote and austere bases; from sea-bases; and across strategic and operational distances.
- Rapidly deploy the joint forcible entry force across the global battlespace, with little or no reception, staging, onward movement, and integration RSOI constraints, and transition to immediate employment in the objective area.
- Provide a dynamic planning, tasking, and execution process that supports the force flow and sustainment of the force.
- Seamlessly and rapidly reconstitute or reconfigure joint forcible entry forces and sustain operations.
- Establish additional contingency airfields or ports, or significantly increasing the existing throughput capacity.
- Reduce supply and re-supply demands through weapons systems with increased precision, effectiveness, firepower and reliability.
- Recognize and rapidly apply technological advances that reduce the demand for all classes of supply in order to enhance joint forcible entry operations: e.g., reduce demand on fossil fuels, miniaturization of ordinance, etc.
- Provide what is needed, and when it is needed, to distributed forces through enhanced capabilities such as predictive logistics, reachback, improved throughput systems, and precise delivery systems.
- Rapidly treat, stabilize and evacuate casualties.[54]

The Seabasing JIC states that

Seabasing and joint logistics are closely aligned. Sustaining joint forces is one of the major seabasing lines of operations, and several key capabilities are shared by both concepts, including C2 [command and control], total asset/in-transit visibility, selective off-load/on-load, and medical and tailored logistical packages. Seabasing provides a viable means . . . from the sea . . . to support logistics for joint forces ashore.[55]

---

[54] DoD, 2004f, p. 43.

[55] DoD, *Seabasing Joint Integrating Concept*, Vers. 1.0, August 1, 2005b, p. 15.

The Seabasing JIC also identifies the following tasks that, although they are not directly related to this CBA, are potentially relevant (Tasks S4–S7):

- Provide continual sustainment to selected joint forces ashore
- Provide personnel and personnel support
- Provide joint maintenance support
- Provide joint medical support.[56]

As previously stated, this CBA does not have a JIC. We will therefore build on the applicable documents to develop the tasks required for this CBA.

The U.S. Air Force Transformation Flight Plan 2004

This "flight plan" shows how ongoing and planned USAF transformation efforts are addressing the Secretary of Defense's transformation planning guidance.[57] The USAF transformation strategy is to

- enhance joint and coalition warfighting
- aggressively pursue innovation
- create flexible, agile organizations
- implement capabilities- and effects-based planning and programming
- develop "transformational" capabilities
- break out of industrial-age business processes.[58]

Several sections of *The U.S. Air Force Transformation Flight Plan* are relevant to intratheater airlift. In particular, the Global Mobility

---

[56] DoD, 2005b, Annex C, Excel Spreadsheet, "Sustain" tab, information dated January 10, 2005.

[57] Headquarters (HQ) USAF, Future Concepts and Transformation Division, *The U.S. Air Force Transformation Flight Plan 2004*, 2004, p. i.

[58] HQ USAF, 2004, pp. ii–iii.

CONOPS section supports the transformation to a capabilities-based force and expresses the desire for rapid projection and application of joint U.S. military power. Two mission areas are especially important to intratheater airlift:

- Power Projection through Air Mobility
    - The seamless integration and effective conduct of air mobility operations in CONUS [the continental United States], en route, or forward locations and with all theater operations.
    - Air Mobility Forces that have the capabilities to seamlessly integrate with joint and coalition forces across all theater boundaries in order to rapidly accomplish the objectives of the combatant commander.
    - The assured ability to deploy, replenish, sustain, and redeploy joint forces in minimum time to allow them to accomplish the missions assigned to them through all phases of conflict.

        . . .

- Power Projection through Expeditionary Air Bases
    - Assured ability to mesh seamlessly with other forces (Army, Marine Corps, and SOF) to open a base and establish air operations from a spectrum of airfields—austere base, cold base, warm base, and hot base (includes CBRNE environments).
    - Achieving seamless transition from airfield seizure to base opening, to force employment and sustainment in concert with theater-assigned mobility forces; includes the rapid, efficient redeployment of forces.[59]

The flight plan states that

Rapid establishment of air operations, an air-bridge, and movement of military capability in support of operations anywhere in the world under any conditions[60]

---

[59] HQ USAF, 2004, pp. 42–43.

[60] HQ USAF, 2004, p. 63.

is a transformational capability and is required to achieve rapid global mobility. As part of Rapid Global Mobility, the USAF has established assessment teams,

> which assess forward airfields in a theater of operations . . . [allowing] seamless integration between airfield seizure and operations . . . [thereby] enhancing the combat power available to the joint force commander. . . . The Air Force is developing a new concept for a specialized unit to rapidly open airfields.[61]

Finally, in "Transforming How the Air Force Does Business," the transformation flight plan identifies important aspects of sustainment transformation and states that "[c]ombat efficiency places a great reliance on the sustainment infrastructure and its business processes."[62] These initiatives could affect intratheater airlift by increasing the availability of the airlift fleet and by improving the situational awareness of airlift operations and cargo movement.

Global Mobility CONOPS

According to the Global Mobility CONOPS, global mobility must provide "rapid projection and application of joint U.S. military power" and "rapid, precise, and persistent delivery" for the joint warfighter.[63] This CONOPS, in conjunction with the other six USAF CONOPS, is the starting point for the USAF capabilities-based planning process and provides the "operational context and high-level capabilities to support global mobility capabilities-based planning."[64]

The CONOPS states that "[g]lobal mobility achieves effects through the application of airlift, air refueling, expeditionary air mobility, spacelift, and Special Operations Forces (SOF) mobility

---

[61] HQ USAF, 2004, p. 65.

[62] HQ USAF, 2004, p. 75.

[63] Much of the text in this section is taken directly from USAF, *Global Mobility CONOPS*, Vers. 4.3, working draft, December 29, 2005, p. 10.

[64] USAF, 2005, p. 2.

capabilities."[65] Although all these may have implications for intratheater airlift, three are of primary importance: airlift, expeditionary air mobility, and SOF mobility.

Airlift provides the ability to deliver combat personnel, their equipment, and their supplies in direct support of combat operations by airdrop to a precisely designated location or by airland operations at established or austere landing zones. Airlift includes aeromedical evacuation and distinguished visitor support.[66]

Expeditionary air mobility operations provides "the ability to rapidly establish, expand, sustain, and coordinate air mobility operations worldwide. . . . from fixed established sites and austere operating areas."[67]

SOF mobility provides "rapid, global airlift of personnel and materiel through hostile or politically sensitive airspace to conduct special operations . . . ."[68]

The Air Mobility Master Plan

The Air Mobility Master Plan (AMMP),[69] is a strategic plan looking out 25 years to guide research and development efforts necessary for developing the capabilities that MAF will need to provide in future operating environments. AMC produces the AMMP, a MAF plan written by MAF members,[70] as the USAF lead command for air mobil-

---

[65] USAF, 2005, p. 3.

[66] USAF, 2005, p. 11.

[67] USAF, 2005, p. 12.

[68] USAF, 2005, p. 15.

[69] Author telephone conversation and email exchange with John Orlovsky, HQ AMC/A55PL, February 27, 2006.

[70] MAF members include the following major commands, reserve components, and combatant command air forces: AMC (lead), Air Combat Command, Air Force Materiel Command, Air Education and Training Command, Air Force Space Command, Air Force Special Operations Command, U.S. Air Forces Europe, Pacific Air Forces, Air Force Reserve Command, Air National Guard, U.S. Central Air Forces (9th Air Force), and U.S. Southern Command Air Forces (12th Air Force).

ity operations. The AMMP is a capabilities-based plan that supports effects-based operations.

The Global Mobility CONOPS is one of several key drivers for the AMMP. It identifies the mobility capabilities that MAF must provide today and in the future. The AMMP is based on an iterative, three-phased strategies-to-tasks planning process. After assessing the future operating environment and guiding documents, teams determine required MAF capabilities and compare them with current mobility capabilities to identify deficiencies. Teams then identify solution sets for the deficiencies. AMMP Roadmaps present MAF capabilities, deficiencies, and solution sets for mission areas, weapon systems, and support processes.

Army Vision

The Army's fixed-wing FAA report discusses relevant Army guidance documents.[71] A few points do specifically affect USAF intratheater airlift. According to these documents,

> Responsiveness has the quality of time, distance and sustained momentum, and implies the inherent capability for preemptive, not just reactive, employment, to influence and shape the outcome of a crisis.

Sustainability is defined as "the capability to continue operations longer than any adversary." The needs for responsiveness and sustainability drive the requirement for intratheater airlift in the battlespace to support the future joint land force.

---

[71] U.S. Army Aviation Center, 2003a, pp. 10–12.

CHAPTER THREE

Operational Mission Areas

As stated previously, much intratheater airlift will be in support of joint land forces. The Objective Force (OF)

> is the Army's future full-spectrum force that will be organized, manned, equipped and trained to be more strategically responsive, deployable, agile, versatile, lethal, survivable and sustainable than we are today—across the full spectrum of military operations as an integral member of a cohesive joint team.[1]

Drawing once again on the Army's Future Combat Aircraft FAA to gain insight into this environment,

> OF full-spectrum operations will be characterized by multiple, concurrent, geographically separated, highly focused operations, executed by tailored Joint and combined-arms, air-ground teams, within a specified theater of operations. OF units will conduct operational maneuver from strategic distances; deploy through multiple, unimproved points of entry/transition, forcibly if necessary; overwhelm hostile, anti-access capabilities and rapidly impose their will upon the enemy.[2]

---

[1] U.S. Army, *The 2003 United States Army Posture Statement*, 2003.

[2] U.S. Army Aviation Center, 2003a, p. 28.

That document further states that "the logistical support centers may be positioned well outside the tactical area, and perhaps, outside the operational scope of the Joint Operations Area (JOA)."[3]

The USAF identified three broad mission areas related to intratheater airlift during the initial discussion on the scope of this project—providing routine sustainment, TS/MC resupply, and maneuver to units of action across all operating environments. During the evaluation of the guidance documents, we found that they also provided a good way to bound and organize the problem. As a result, these three mission areas serve as the collectors of tasks and provide the analytical construct for this CBA.

Routine sustainment is defined as the steady-state logistical flow of required supplies and personnel to deployed units. The consumption rate for many items is generally well understood, so the required routine sustainment can be identified and planned well in advance. These items may consist of water, food, and other items needed to conduct planned operations.

Fuel and water will typically constitute the majority of routine sustainment (about 60 to 85 percent).[4] The predictable nature of this requirement will allow for preplanned airlift operations and efficiently loaded airlift sorties. The ability of the intratheater airlift system to fulfill this requirement is driven by the quantity of supplies and the number of personnel that must be moved by air over time and the number of delivery locations that must be supported.

The capability to provide *TS/MC resupply* is generally reflected by the ability of the airlift system to respond to short-turn taskings for crucial equipment, supplies, and personnel. The requirement for this capability is driven by the need for items with unpredictable consumption rates and the need for items that are not kept on hand at every operational location. This requirement will primarily be in the form of

---

[3]   U.S. Army Aviation Center, 2003a, p. 28.

[4]   This range results from differences in unit type, amount of maneuver, and level of combat. Heavy units require more fuel than lighter units. Assumptions on amount of maneuver and level of combat (ammunition expenditure) also affect the percentage of liquid as opposed to bulk sustainment.

demands for delivery of spare parts required to keep equipment operational or emergency supplies of ammunition or other critical items that were expended faster than predicted (e.g., fuel). TS/MC resupply may also include delivery of key personnel with specific skills—perhaps personnel required for equipment repair or to conduct a particular task. The ability of the intratheater airlift system to fulfill this requirement is driven by the quantity of supplies required over time, the number of personnel, the number of delivery locations that must be supported, and the acceptable delivery time.[5]

Maneuver is defined as the ability of the intratheater airlift system to transport combat teams around the battlefield. The maneuver task is associated with the initial deployment, redeployment, and extraction of these teams as required. Maneuver missions may include (but are not limited to)

- transport to mission locations prior to commencement of mission
- transport to mission in progress
- transport from one mission area to another
- transport following completion of a mission to include moving the mission team as well as any materiel or personnel acquired during operations (i.e., rescued personnel or captured enemy or materiel).

The ability of the intratheater airlift system to fulfill the maneuver requirement is driven by the number of teams that must be moved over time; the size of the teams, including the required equipment; and the location to which these teams must be delivered. These combat teams' supply requirements during ongoing operations may fall into either the routine sustainment or TS/MC resupply task.

---

[5]   The phrase "acceptable delivery time" addresses the need to meet short-turn taskings. This can drive the airlift requirement if many separate airlift missions must be flown to meet very short delivery time lines.

CHAPTER FOUR

Scenarios and Operational Environment

To conduct a robust CBA, it is important to choose a variety of scenarios and operational environments. Scenarios can drive both the amount of force structure required and the required military capabilities and potential shortfalls.

The primary focus of the analysis of potential future conflicts is the Defense Planning Scenarios. These scenarios provide the approved requirement on which to base force structure decisions. As a result, this is a required element of the JCIDS process.

In addition, other scenarios and historical operations can be used to understand the issues more fully. The recent operations in Afghanistan and Iraq have supplied a vast amount of real-world operational experience with the intratheater airlift system. Although it is important to understand issues, scenarios, and environments that are not part of the accepted scenarios, these non–JROC-approved excursions may not be credible in the DoD acquisition process.

Planning Scenarios

The Defense Planning Scenarios identify a broad range of challenges that must be considered. We will consider and evaluate aspects important to the intratheater airlift mission. The Multiservice Force Deployment and Operational Analysis 2006 will help us identify potential differences from past major combat operations and historical cases. In addition, the FNA and the FSA will incorporate applicable results and analysis from the Mobility Capabilities Study.

Historical Experience

Recent experience in Afghanistan and Iraq provides insight into the current operational environment and operational approach. Although one should be careful about relying too heavily on an historical experience so as not to "fight the last war," these historical cases, used correctly, provide enormous insight into supply and maneuver operations. Further, it is important to understand these operations because they are arguably the catalyst for the perceived shortfalls that led to the Army's CBA, as well as this CBA. Therefore, an understanding of the capabilities of the intratheater airlift system using historical cases is a required foundation for the FNA and the FSA.

Past operations provide good insight into the consumption rates for supplies and the basis for understanding TS/MC resupply. These operations also provide the only real source for quantities to be delivered, when, and where. These operations, however, may reflect only a portion of the operational environment in which the future intratheater airlift system must operate.

The operations in Afghanistan have had several distinct characteristics that must be considered. First, Afghanistan is a highly mountainous region. Operations often take place at relatively high altitudes that adversely impact aircraft performance—especially the takeoffs of helicopters and fixed-wing aircraft. In addition, Afghanistan is very far from bases in the continental United States and those overseas with permanent U.S. forces. Finally, Afghanistan is landlocked, and the political situation required great dependence on airlift for transport of forces and supplies.[1] All these factors greatly influenced the development of CONOPS for providing routine sustainment.

The experience in Iraq has also had unique characteristics. First, perhaps the most striking aspect of Iraq is the desert environment. The flat, bald terrain could lead to demands for "niche capabilities" that would not be suitable or usable in forested or mountainous regions. Contrary to the experience in Afghanistan, a dozen or more very well

---

[1]   Even the Marine Expeditionary Unit that landed on the coast of Pakistan flew over parts of Pakistan instead of road-marching the entire distance.

prepared bases exist on Iraq's doorstep. Further, the United States has been operating from many of these bases for over a decade, since the deployments during Operation Desert Storm.

Critical Variables of the Operational Environment

As discussed earlier, the assumptions about the operational environment can have a major influence on the analytical results. Care must be taken to evaluate a range of potential environments to ensure a robust solution. Important characteristics include the amount of supplies that must be delivered, the criticality of the timing of that delivery, the degree to which the demand can be anticipated, the number and size of offload points, the nature and condition of the supply base, the threat environment, and the infrastructure of the delivery base.

A critical aspect of the environment is the location where the cargo needs to be delivered. This aspect is classically discussed in terms of airbase infrastructure. The length of the runway, pavement strength, apron parking space, fuel storage, and other aspects greatly affect the ability to conduct operations. These factors are important because they affect airlift platform capabilities in different ways. A C-5, for example, requires a much larger, more-developed airfield than does a C-130, which can operate from fairly austere airfields. Even if two aircraft can operate from a short runway, the payload can be different. In some cases, a short runway could constrain the payload of a particular aircraft due to landing distance while another aircraft that has better breaking could deliver its maximum payload to that base. In all cases, the analysis will use consistent measures of takeoff performance—such as critical field length—and a variety of takeoff environmental conditions to ensure robust and consistent results. Similarly, it will also use consistent measures of landing distance.

In addition, the number of air bases and the desired CONOPS can determine the suitability of particular platforms and can affect the analysis of potential shortfalls. In the case of delivery of personnel and supplies to an open field, the unobstructed length and condition of the field will drive the capability of particular platforms to meet the requirement. In these cases, the ability of aircraft to operate from soft and unimproved fields must be considered. Further, soft-field performance is usually measured in terms of the number of passes that specific aircraft at particular weights can make before the field is rutted-out beyond use. Comparisons that consider the amount of cargo that can be delivered over time before the field is unusable are important.

Another issue that may be important is the threat to fixed- and rotary-wing aircraft. Increasingly lethal ground-based air-defense systems have been proliferating over the last several years. Man-portable air defense systems (MANPADS) could be a significant problem. MANPADS use passive sensor technologies and therefore provide no warning of an attack until the missile is launched. MANPADS are also easy to move, hide, and operate, further complicating the problem because an air defense threat could potentially "appear" nearly anywhere in the battlespace. On the other end of the ground-based air defense spectrum are radar-guided surface-to-air missiles. These long-range systems could be a significant challenge to certain missions that may operate in a higher threat environment—notably SOF maneuver.

Analytical results will also be sensitive to variables describing the operational environment and operational tasks. These variables have a great deal of influence on the capabilities required to conduct the tasks. Table 4.1 lists some of the more important variables. An evaluation of these variables is required to analyze the potential shortfalls in current USAF capability.

Table 4.1
Variables Describing Operational Environment

	Variable	Routine Sustainment	Time-Sensitive, Mission-Critical Resupply	Small-Unit Maneuver
Delivery environment and insertion method	Number of delivery points	X	X	X
	Accessibility of lines of communication	X	X	X
	Proximity of airfield, port, or waterway	X	X	X
	Terrain	X	X	X
	Threat level	X	X	X
Physical characteristics of supplies	Class of supply	X	X	X
	Volume	X	X	X
	Weight	X	X	X
	Total amount	X	X	X
Team characteristics	Number of people	X	X	X
	Amount and type of equipment	X	X	X
Mission type	Level of covertness			X
	Personnel or equipment to pick up			X
	– Snatch and grab			X
	– Hostage or casualty evacuation			X
	– Recovery of weapons of mass destruction			X

CHAPTER FIVE

Tasks

Table 5.1 identifies the UJTL tasks that are relevant to each mission area being considered in this intratheater airlift CBA. Table 5.2 presents the relevant tasks for the Air Force MCL.

The Air Force Task List is derived from the current Air Force MCL.[1] Only capabilities from the MCL will be included in this document.

---

[1] Telephone conversation with Lt Col Brent Phillips, HQ USAF/XOOA, February 22, 2006.

Table 5.1
Universal Joint Task List

		Mission Areas		
Universal Joint Tasks		Routine Sustainment	Time-Sensitive, Mission-Critical Resupply	Small-Unit Maneuver
OP 1.1	Conduct operational movement			X
OP 1.1.2	Conduct intratheater deployment and redeployment			X
OP 1.1.2.1	Conduct airlift in the joint operations area	X	X	X
OP 1.1.3	Conduct joint reception, staging onward movement, and integration in the joint operations area			
OP 1.2	Conduct operational maneuver and force positioning			X
OP 1.3	Provide operational mobility			X
OP 1.3.2	Enhance movement of operational forces	X	X	X
OP 1.6	Conduct patient evacuation	X	X	X
OP 4	Provide operational logistics and personnel support	X	X	X
OP 4.3	Provide for maintenance of equipment in the joint operations area	X	X	
OP 4.4.3	Provide for health services in the joint operations area	X	X	X
OP 4.5.1	Provide for movement services in the joint operations area	X		X

Table 5.1—Continued

Universal Joint Tasks		Mission Areas		
		Routine Sustainment	Time-Sensitive, Mission-Critical Resupply	Small-Unit Maneuver
OP 4.5.2	Supply operational forces	X	X	
OP 4.7	Provide politicomilitary support to other nations, groups, and government agencies	X	X	X
OP 4.7.2	Conduct civil military affairs	X	X	X
OP 4.7.3	Provide support to DoD and other government agencies	X	X	X
OP 5.5.4	Deploy joint force headquarters advance element			
OP 6.2	Provide protection for operational forces, means and noncombatants			
OP 6.2.6	Conduct evacuation of noncombatants from the joint operation area			X
TA 1	Deploy/conduct maneuver			X
TA 1.1.1	Conduct tactical airlift	X	X	X
TA 1.2.4	Conduct counterdrug operations	X	X	X

Table 5.2
Air Force Master Capabilities Library

		Mission Application	
Air Force Capabilities	**Routine Sustainment**	**Time-Sensitive, Mission-Critical Resupply**	**Small-Unit Maneuver**
6.1.1.1 Airlift Materiel	X	X	X
6.1.1.2 Airlift Personnel	X	X	X
6.1.1.3 Airlift Distinguished Visitors	X	X	X
6.1.1.4 Airlift Patients (Aeromedical Evacuation)	X	X	X
6.1.1.5 Infiltrate/Exfiltrate Special Operations Forces/Battlefield Airmen and Materiel	X	X	X

CHAPTER SIX

Concluding Remarks: Tasks, Conditions, and Standards

The objective of this CBA is to analyze a potential deficiency in intratheater airlift capability: The demands of the global war on terrorism and new operational requirements may create shortfalls in the ability of USAF intratheater airlift to deliver personnel and equipment. The USAF identified three broad operational mission areas for this evaluation of the intratheater airlift system. These are its ability to provide

1. routine sustainment
2. TS/MC resupply
3. maneuver

to U.S. and allied forces across all operating environments.

Working with the relevant guidance documents (see Chapter Two) and the UJTL and the MCL (see Chapter Five), we developed the tasks, conditions, and standards that are important for this CBA. Table 6.1 presents these tasks and identifies how they apply to each of the three mission areas for this CBA.

Although the guidance documents do not specify a set of conditions under which these tasks *must* be accomplished, the attributes and conditions are discussed throughout the guidance documents. Some of these attributes and/or conditions occur in multiple guidance documents. We identified the following conditions that the Air Force identified as important and should be considered in this CBA:

Table 6.1
Tasks and Mission Areas Applicable to This CBA

Task	Routine Sustainment	Time-Sensitive, Mission-Critical Resupply	Small-Unit Maneuver
Transport supplies and equipment to points of need	X	X	X
Conduct retrograde of supplies and equipment	X	X	X
Transport forces and accompanying supplies to point of need[a]			X
Recover personnel and supplies[b]			X
Transport replacement and augmentation personnel	X	X	X
Evacuate casualties	X	X	X

[a] Deployment, redeployment, and retrograde.

[b] Includes evacuation of hostages, evacuees, enemy personnel, and high-value items.

- Adverse weather.
- Multiple, simultaneous, distributed decentralized battles and campaigns.
- Anti-access environment.
- Support forces operating in and from austere or unimproved locations.
- Degraded environments (WMD/WME, CBRNE, Natural disasters).
- Multi-national environment.[1]
- Absence of pre-existing arrangement.

The following attributes and conditions are discussed as positive in the guidance documents:

---

[1] The *multinational environment* includes both support to non–U.S. forces in situations in which U.S. ground forces are not engaged at all or have limited involvement as advisers and support of friendly and allied forces with and without the participation of U.S. ground forces.

- establishing the smallest logistical footprint
- delivering with speed, accuracy, and efficiency
- distributing to the point of requirement
- basing flexibly to permit operation across strategic and operational distances.

The guidance documents also specify standards that should be used to evaluate potential gaps in capabilities. The tasks identified above should be accomplished with the following standard capabilities in mind:

- meeting demands for force and materiel movement
- moving forces and materiel throughout a theater optimally
- providing materiel support for current and planned operations.

A critical aspect of the environment is the location to which the cargo is to be delivered. Another issue is the threat to fixed- and rotary-wing aircraft. Increasingly lethal ground-based air-defense systems have been proliferating over the last several years. This threat ranges from MANPADS on the low end to highly capable long-range surface-to-air missiles on the high end.

Analytical results will also be sensitive to variables describing the operational environment and operational tasks. These variables have a great deal of influence on the capabilities required to conduct the tasks. Some of the more-important variables are number of delivery points; terrain; airbase accessibility; total amount, dimensions,[2] and weight of each supply class required to be delivered by air; number of personnel required to be delivered by air; required response time; and threat level.

---

[2] The length, width, and height of the cargo compartment are critical for determining what can be transported. The dimensions of the cargo compartment and payload capability will drive the design of ground combat vehicles for as long as an aircraft is projected to be a significant portion of the airlift fleet.

Bibliography

Air Mobility Command, U.S. Air Force, *Air Mobility Master Plan, AMC/A55PL*, October 2004.

Bush, George W., *The National Security Strategy of the United States of America*, Washington, D.C.: The White House, September 17, 2002.

Chairman of the Joint Chiefs of Staff, Instruction (CJCSI) 3170.01E, *Joint Capabilities Integration and Development System*, May 11, 2005.

————, CJCSI 3170.01F, *Joint Capabilities Integration and Development System*, May 1, 2007.

————, CJCSI 3170.01G, *Joint Capabilities Integration and Development System*, March 1, 2009.

————, Manual (CJCSM) 3170.01C, *Operation of the Joint Capabilities Integration and Development System*, May 1, 2007.

CJCSI—*See* Chairman of the Joint Chiefs of Staff, Instruction.

Department of Defense Instruction 5000.2, *Operation of the Defense Acquisition System*, May 12, 2003.

————, *Operation of the Defense Acquisition System*, December 8, 2008.

DoD—*See* U.S. Department of Defense.

Headquarters U.S. Air Force, Future Concepts and Transformation Division, *The U.S. Air Force Transformation Flight Plan 2004*, Washington, D.C., January 1, 2004.

HQ USAF—*See* Headquarters U.S. Air Force.

JCS—*See* Joint Chiefs of Staff.

Joint Chiefs of Staff, *The National Military Strategy of the United States of America*, Washington, D.C., 2004a.

————, *Focused Logistics Campaign Plan*, 2004b.

———, *Joint Logistics (Distribution) Joint Integrating Concept*, Vers. 1.0, February 7, 2006.

Joint Staff, Force Application Assessment Division (J8), "Conducting a Capabilities-Based Assessment (CBA) Under the Joint Capabilities Integration and Development System (JCIDS)," white paper, January 2006.

———, *Joint Capabilities Integration and Development System (JCIDS)*, January 2006.

Joint Staff, Joint Experimentation, Transformation, and Concepts Division (J7), *Capstone Concept for Joint Operations*, Vers. 2.0, August 2005a.

———, "JOpsC Family of Joint Concepts—Executive Summaries," briefing, August 23, 2005b.

———, "Joint Integrating Concepts," *Future Joint Warfare* website, last update January 2, 2008. As of February 7, 2008:
http://www.dtic.mil/futurejointwarfare/jic.htm

Joint Staff (J7)—*See* Joint Staff, Joint Experimentation, Transformation and Concepts Division.

Orletsky, David T., Daniel M. Norton, Anthony D. Rosello, William Stanley, Michael Kennedy, Michael Boito, Brian G. Chow, and Yool Kim, *Intratheater Airlift Functional Solution Analysis (FSA)*, Santa Monica, Calif.: RAND Corporation, MG-818-AF, 2011. As of February 3, 2011:
http://www.rand.org/pubs/monographs/MG818.html

Stillion, John, David T. Orletsky, and Anthony D. Rosello, *Intratheater Airlift Functional Needs Analysis (FNA)*, Santa Monica, Calif.: RAND Corporation, MG-822-AF, 2011. As of February 3, 2011:
http://www.rand.org/pubs/monographs/MG822.html

USAF—*See* U.S. Air Force.

U.S. Air Force, *Global Mobility CONOPS*, Vers. 4.3, working draft, December 29, 2005.

U.S. Army, *The 2003 United States Army Posture Statement*, 2003. As of February 7, 2008:
http://www.army.mil/aps/2003/realizing/transformation/operational/objective/index.html

U.S. Army Aviation Center, Futures Development Division, Directorate of Combat Developments, *Army Fixed Wing Aviation Functional Area Analysis Report*, Fort Rucker, Ala., June 3, 2003a.

———, *Army Fixed Wing Aviation Functional Needs Analysis Report*, Fort Rucker, Ala., June 23, 2003b.

———, *Army Fixed Wing Aviation Functional Solution Analysis Report*, Fort Rucker, Ala., June 8, 2004.

U.S. Army, Training and Doctrine Command Analysis Center, *Future Cargo Aircraft (FCA) Analysis of Alternatives (AoA)*, July 18, 2005, not releasable to the general public.

U.S. Army and U.S. Air Force, "Way Ahead for Convergence of Complementary Capabilities," memorandum of understanding, February 2006.

U.S. Department of Defense, *Joint Operations Concepts*, November 2003a.

———, *Focused Logistics Joint Functional Concept*, Vers. 1.0, Washington, D.C., December 2003b.

———, *Homeland Security Joint Operating Concept*, Washington, D.C., February 2004a.

———, *Strategic Deterrence Joint Operating Concept*, Washington, D.C., February 2004b.

———, *Force Application Functional Concept*, Washington, D.C., March 5, 2004c.

———, *Stability Operations Joint Operating Concept*, Washington, D.C., September 2004d.

———, *Major Combat Operations Joint Operating Concept*, Washington, D.C., September 2004e.

———, *Joint Forcible Entry Operations Joint Integrating Concept*, Vers. 92A3, limited distribution draft, Washington, D.C., September 15, 2004f.

———, *The National Defense Strategy of the United States of America*, Washington, D.C., March 2005a.

———, *Seabasing Joint Integrating Concept*, Vers. 1.0, Washington, D.C., August 1, 2005b.

———, *Homeland Defense and Civil Support Joint Operating Concept*, draft, Vers. 1.5, November 2005c.

———, *Quadrennial Defense Review Report*, Washington, D.C., February 6, 2006.

———, *The National Defense Strategy of the United States of America*, Washington, D.C., June 2008.

Yarnell, Edward, "Joint Transformation Concepts," briefing, Joint Experimentation, Transformation and Concepts Division (J7), January 6, 2006.